About the front cover:

A statue, like the image on the cover, was commissioned in 1914 by AT&T's president, sculpted by Evelyn Beatrice Longman, and hoisted to the roof of AT&T Corporate Headquarters in Lower Manhattan. This towering figure became New York's second-largest sculpture, after the *Statue of Liberty*. The gilded *Spirit of Communication* had earlier names, but is most commonly referred to as *Golden Boy*. Before the arrival of cell phones, it was common practice for AT&T's local Bell Telephone companies to print directories of subscriber's phone numbers, and the image on the cover was often used on the cover of those directories.

← *Golden Boy*

Golden Boy statue, previously on top of AT&T headquarters

About the book:

How in the world can sounds be sent from here to there, even without a wire? Written by a physicist for his adult grandchildren, this little book explains the concepts, and it's too good not to share. Three early discoveries in electricity & magnetism led directly to the telegraph, telephone, and radio, respectively. Their mysteries are described here without math, and only the bare essentials are covered – the rest is just details. It will be eye-opening if your interests include technology's history, STEM education, amateur radio, or how things work in general. It may even be helpful to students who plow through complicated mathematics without a visceral feeling for these concepts and their history. The book is beautifully illustrated with color photographs and diagrams.

Naked Physics

of the
Telegraph, Telephone, and Radio

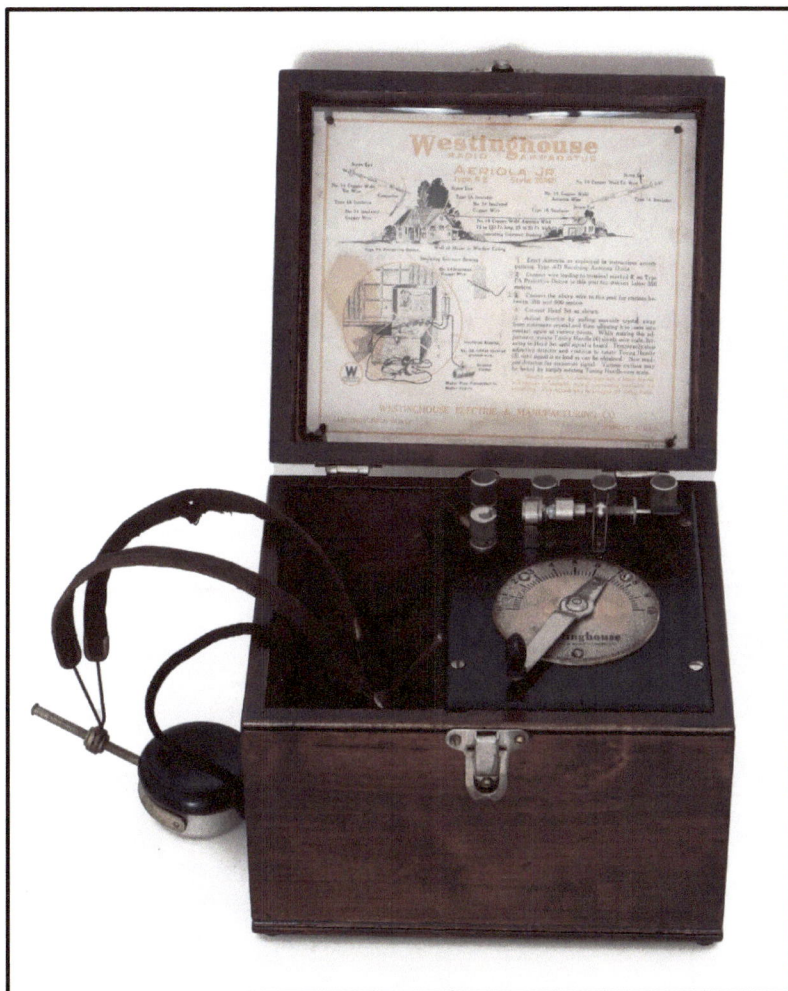

1921 *Aeriola Jr.*, first radio produced for broadcast listening.

Ralph O. Meyer

By the same author:

Old-Time Telephones: History, Design, Technology, Restoration
3rd Edition, North Carolina State University, Raleigh (2018)

Henry Dreyfuss: Designing for People (co-author)
State University of New York Press, Albany (2022)

Henry Dreyfuss and the Telephone: Origins of Industrial Design
(working title, in preparation with co-author)

For all my grandchildren:
a challenge

Naked physics / by Ralph O. Meyer.
Copyright © 2022 by Ralph O. Meyer
ISBN 979-8-218-10421-4

Contents

Ralph O. Meyer

Foreword

During the summer in between my sophomore and junior years in physics at BYU, I was working full time during the day in my dad's physics lab at Denver University and working as a waitress in the evening to make money for college. The waitressing job was really my favorite – I loved it and loved the people I got to meet every time I was in the restaurant.

One evening I was waiting on a large table of business men in suits who were having very serious technical discussions. I listened for a bit then asked them who they worked for, just trying to participate in their little chat. One man was quite rude and obnoxious and basically said, "You are just a dumb waitress – what would you know about these things anyways?" and carried on with his discussion.

Well that TOTALLY ticked me off, so I walked off to the side, grabbed a bar napkin and wrote down Maxwell's equations (since that was the most recent thing we had studied right before the summer break and the first nerdy thing that came to mind). I went off to the bar to get the men's drinks, then when I returned, I passed out the drinks, nonchalantly tossed the Maxwell napkin on the table, and walked off.

I hid around the corner to see their reaction and it was pretty hilarious. One guy grabbed the napkin, looked at it with disbelief in his eyes, then they all started babbling and looking around the restaurant for me. I waited a bit, then took their entrees out to them, and they all appeared a bit flabbergasted. Someone asked me what kind of calculator I used; I said Hewlett Packard. Another guy asked what kind of test equipment I used; I replied Hewlett Packard (as that was all my dad had in his lab!). Someone else asked who made the best technical computers; I assured them it was Hewlett Packard.

They were all beaming and looking pretty pleased with themselves, but I still had no idea who they were. The original guy who was such a jerk pulled a business card out of his pocket and handed it to me. He was the VP for Hewlett Packard in all of the Americas. He asked me who the heck I was (as if I were a freak of nature!) and what I was doing there. I told them I was a physics student but was home working for the summer, thus the waitress work.

He offered me a job on the spot, and I was still a little bit hurt about how they had assumed I was an idiot just because I was a waitress, so I played pretty hard to get. I told him maybe or maybe not, but we could keep in touch. We did keep in touch, and by the next summer he had convinced me, so I worked for him in the Denver office the summer between my junior and senior years. We became very good friends, and he is the one that helped set up my job in DC, as I told him I really wanted to work on the east coast. And he assured me that he has never judged a waitress again!!

Kerri C. Meyer
Daughter-in-law

Introduction

In the movie *IQ*, Albert Einstein, in his later years as a physics professor at Princeton, is played by Walter Matthau. He holds the compass his father gave him and wonders "by what force, invisible and unfelt, could be holding the needle?" Later he says, "It's my compass what got me started in my work." The strange force was a magnetic field.

There are three force fields we are familiar with: magnetic fields that hold a compass needle, electric fields that push electrons around making current for light bulbs, and gravitational fields that cause apples to fall to the ground. These fields all exist in space and have directions and magnitudes. Each can be caused by a variety of sources, and each gets weaker as the distance from its source increases. Fields of the same kind from several sources can add together; e.g., the gravitational field of the earth keeps water in the oceans, and the gravitational field of the moon produces tides.

When I was 13, I read a book about electricity and learned how to make an electromagnet. I wound coils of wire around nails, cut metal strips from tin cans for contacts, and strung a telegraph line between our attic and a tree house. Later while in graduate school, I built a tunable converter that would let me listen to aircraft radio chatter (Cleared for takeoff.) through the radio in our Volkswagen. I knew quite a bit about electricity & magnetism, but I didn't really understand radio waves – and especially how a wire antenna could be part of a circuit when one end of it wasn't connected to anything – or was it? In my old age, I got help to understand these things and then wrote this book for my grandchildren as well as myself, I guess. And my old Boy Scout compass played a role in it.

Electric fields and magnetic fields underlie almost everything that's discussed in this little book, which covers only the bare essentials. Hopefully, these essentials will explain the mysteries of electronic communication. Everything else is just details.

Chapter 1. Hans Christian Ørsted

In 1819, the Danish physicist Hans Christian Ørsted showed that a magnetic needle is deflected by an electric current (i.e., moving electric charges). His experiment is easy to duplicate as shown below. In these pictures, I aligned the wire so it is running north and south (well, almost), and I placed my 1950s Boy Scout compass directly beneath the wire. In Fig. 1, when no current is flowing (light is off), nothing happens; the needle is being held by the earth's magnetic field and points north.

Fig. 1. Compass needle is undisturbed by a wire with no current.

When current is flowing as in Fig. 2 (light is on), a magnetic field adds to the earth's magnetic field and pushes the needle to about 15 degrees (that thing at 65 degrees is a "keeper" to hold the needle still when the compass is being moved). The electric current is creating a magnetic field, and the mathematical relation between the electric current and the magnetic field strength was developed by André-Marie Ampère. Thus this effect is referred to as Ampère's law.

Fig. 2. A magnetic field is created by an electric current.

If the wire is wrapped around itself as in Fig. 3, then more current passes over the compass and the magnetic field is strengthened. In this case it pushes the needle all the way to 60 degrees.

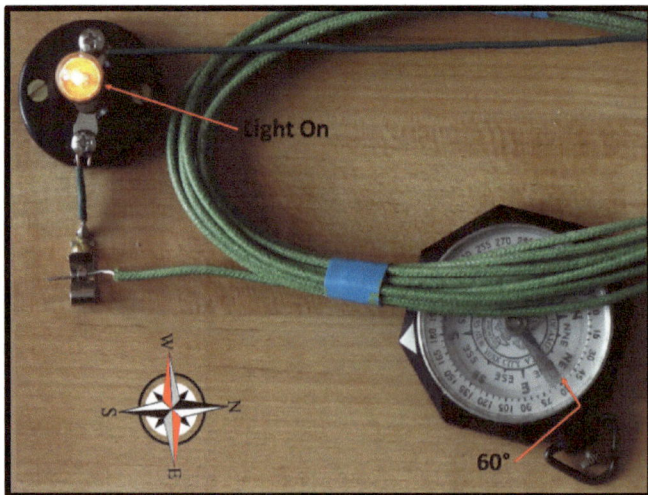

Fig. 3. More electric current gives a stronger magnetic field.

A few comments about magnetic properties of iron are needed to proceed. Atoms in many elements have magnetic fields caused by electrons that are circling their nuclei (i.e., by moving electric charges), and the magnetic field of iron atoms is particularly strong.

These atoms behave like little magnets, but they are normally arranged randomly with no net effect. However, when a piece of iron is placed in an external magnetic field, many of these atomic magnets align themselves, thereby causing the piece of iron to behave like a relatively strong magnet with a north and south pole (i.e., the iron becomes polarized).

In soft iron or mild (bendable) steel, these atomic magnets can change their alignment relatively easily. But some ferrous magnets (so-called permanent magnets) can hold their magnetism almost indefinitely if they are made with hardened steel or special alloys that impede the movement of atomic magnets. Magnetization of these materials is accomplished by using an extremely strong magnetic field to force their atomic magnets into alignment, where they become effectively frozen.

These effects can be seen in Fig. 4. Notice on the left of the figure how the little atomic magnets in the mild steel nail have aligned themselves with the permanent magnet's field such that the nail's south pole is at its tip and its north pole is at its head (opposites attract). When the permanent magnet is flipped over (right side of Fig. 4), the little atomic magnets in the nail obligingly turn around the other way (north pole at the tip and south pole at the head). When the nail is moved away from the permanent magnet, the little atomic magnets return to their random orientation and the nail is no longer magnetized.

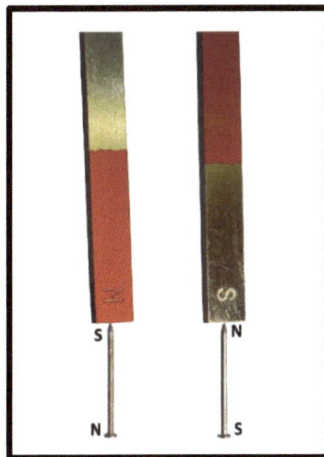

Fig. 4. Alignment of atomic magnets in iron can change easily.

Now if we wrap enamel-insulated wire around a nail to form a coil (Fig. 5), the coil of wire multiplies the number of times current goes past a given point just as in Fig. 3, and the coil's magnetic field aligns many of the little atomic magnets in the nail, again multiplying the magnetic strength. The resulting magnetic field of the nail is now strong enough to pick up a bolt. This arrangement is, of course, an electromagnet.

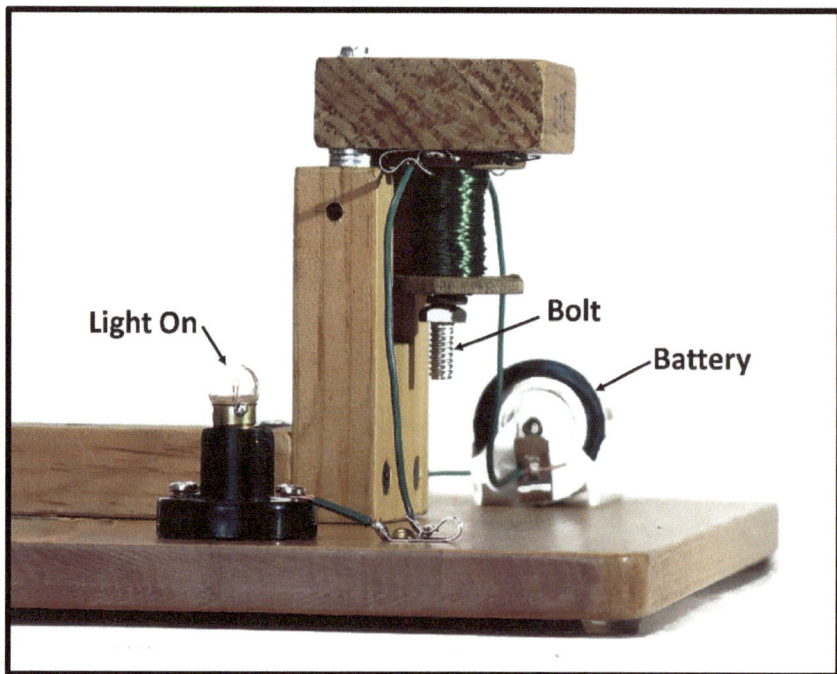

Fig. 5. Current through a coil of wire wrapped around a nail creates a strong magnetic field.

If an alternating current had been used rather than the battery's direct current, the resulting magnetic field would have been alternating (i.e., increasing and decreasing with time). Such an oscillating magnetic field has profound consequences as seen in Chapter 3.

Chapter 2. Telegraph

Before the telegraph was invented in the late 1830s, it could take two weeks to get a message from New York to Chicago – hard to imagine. The first successful telegraph was in fact developed in England by William Cooke and Charles Wheatstone. Like Ørsted's experiments, their instruments used magnetized needles that were deflected by magnetic fields from electric currents in nearby coils.

In the United States, Samuel Morse used an electromagnet instead of deflecting needles. In its early forms, the electromagnet moved a soft-iron part called an armature that carried a pencil or stylus, which in turn left dots and dashes on a scrolling strip of paper (Morse code). Later, the stylus was left off because telegraphers could recognize the code just from the instrument's clicking noise.

A telegraph transmitter, called a key (Fig. 6), was just a switch that would start and stop current going over telegraph wires and then through the receiver's coils.

Fig. 6. A telegraph key is just a switch.

An improved late 19th century telegraph receiver, called a sounder (Figs. 7), used a pair of coils for better performance than a single coil. It had a spring-loaded brass bar (brass is not magnetic like iron) that would hit a lower stop when the electromagnet pulled the iron armature down – and then hit an upper stop when it was released. The sound of the brass bar hitting the stops was heard by skilled telegraphers, who translated the clicks into words on the fly!

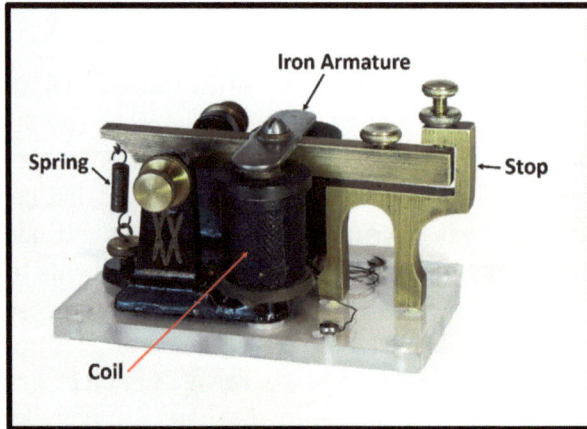

Fig. 7. Typical telegraph sounder of late 1800s.

A complete telegraph system would include a key and battery at telegraph Office A to send a message, and a sounder at Office B to receive the message. A simplified diagram of such a telegraph system (Fig. 8) shows the connections between the key and sounder as well as a required battery. Telegraph wires (lines) often ran along railroad tracks because one of the first big applications of the telegraph was in railroad operation.

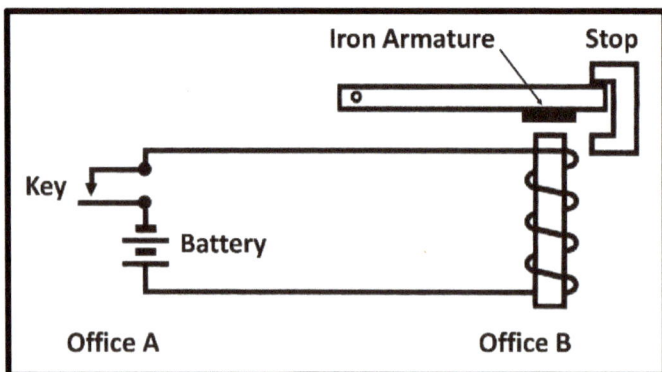

Fig. 8. Simplified diagram of a telegraph system.

Chapter 3. Michael Faraday

In 1831, Michael Faraday, an English scientist, showed that a changing magnetic field would induce an electric current in a nearby wire circuit.[1] The pictures below show a circuit I made by connecting the ends of my coil (same one as in Fig. 5) to a sensitive current meter called a galvanometer. When the round magnets are sitting stationary on the nail head, nothing happens; the needle on the galvanometer is stationary at zero (see Fig. 9). Keep in mind that the magnets magnetize the nail (i.e., polarize it just as in Fig. 4) such that there is a strong magnetic field near the wire that is wrapped around the nail. But the magnetic field is constant, i.e., not changing with time, so it does not generate an electric current.

Fig. 9. A steady magnetic field does not create an electric field.

---

[1] In the same year, the American, Joseph Henry, independently discovered magnetic induction, but the Englishman, Faraday, was first to discover it and publish it. Henry (like Edison) appears to have been more of a tinkerer than a real scientist like Faraday – and Faraday got his name on the mathematical law.

However, if the magnets are moved quickly away from the nail (or toward it), an electric current is generated in the wire. In Fig. 10, my assistant was moving the magnets away from the nail head when I snapped the picture. Her hand was moving (blurred in the photo) and the needle was swinging when I captured this photo with the needle at -28 micro amperes (μA).

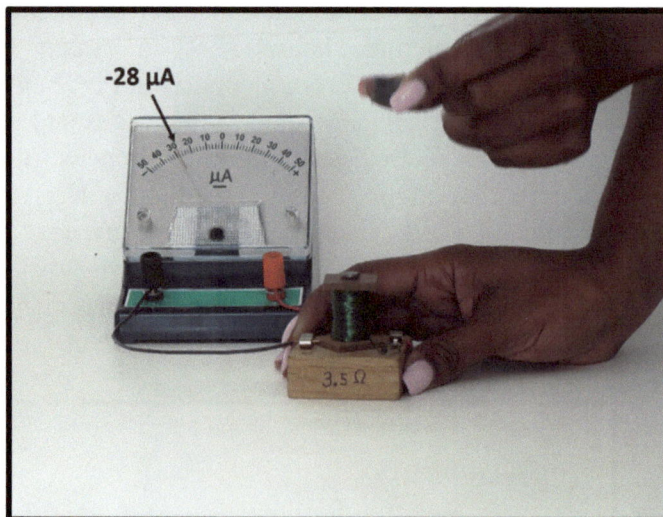

Fig. 10. A changing magnetic field creates an electric field.

Faraday deduced from his observations that a time-varying magnetic field had produced an electric field, which was then able to push charges around in a wire.[2] Electric charges can only be moved by electric fields, not by magnetic fields. The equation between the time-rate-of-change of the magnetic field and the electric field is called Faraday's law.

Now we can begin to see the importance of alternating current. If alternating current were used to create an alternating magnetic field (Ampère's law), then this alternating magnetic field would produce an alternating electric field (Faraday's law). Crazy!

---

[2] Metals such as copper are used in wires, and copper contains electrons that can be pushed around easily by an electric field. Insulators such as porcelain (see back cover) have electrons that can't be pushed around at all by an electric field.

Chapter 4. Telephone

To send a telegram in the mid-1800s, you went to a telegraph office, wrote out your message, gave it to a skilled telegrapher who telegraphed it to another office, where your message was written on paper and delivered by a messenger. A telephone call would be faster, cheaper, and let you hear the other person's voice as well.

Transmitters

Alexander Graham Bell knew about the work of Faraday and Henry, and Bell hypothesized that he could convert pressure waves from vocal sounds to electrical currents by using that pressure to vibrate a magnet in front of the poles of a telegraph coil. Because current is generated this way, Bell's phone was called a magneto telephone, and it became the standard for the first two years of the commercial telephone (1877-1878). The magneto telephone was used as a transmitter and a receiver, had a range of more than 100 miles, and used no batteries. It was scientifically beautiful.

The workings of this telephone (Fig. 11) are a little complex, and it is not clear that Bell understood all of the physics. But you can.

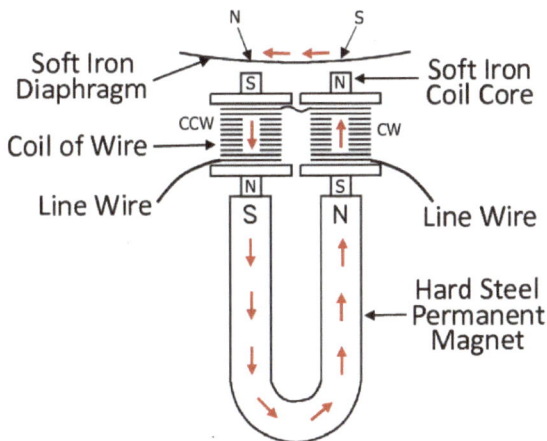

Fig. 11. Functional diagram of Bell's magneto telephone.

First, consider the permanent magnet, whose little atomic magnets are frozen in place by the hardening process. Next there is a soft-iron pin (a coil core) on each pole of the permanent magnet, and the permanent magnet will magnetize these two iron coil cores. Finally there is a thin soft-iron diaphragm very close to the coil cores, and the diaphragm is magnetized by the coil cores. The coil cores and diaphragm are thus magnetized, but their atomic magnets are not frozen and could be changed further.

When sound waves vibrate the diaphragm, it moves toward and then away from the coil cores, alternatively increasing and then decreasing the gap and hence strength of the magnetic field in the coil cores. The changing magnetic field in the coil cores produces an electric field that is felt by the wires, and a current is induced – just as it was in Fig. 10. The generated current has the frequency and amplitude (i.e., the pitch and loudness) of the original sound.

The inside of this so-called box telephone is shown in Fig. 12, and a box was fitted over the top. This replica was made by Bell Labs.

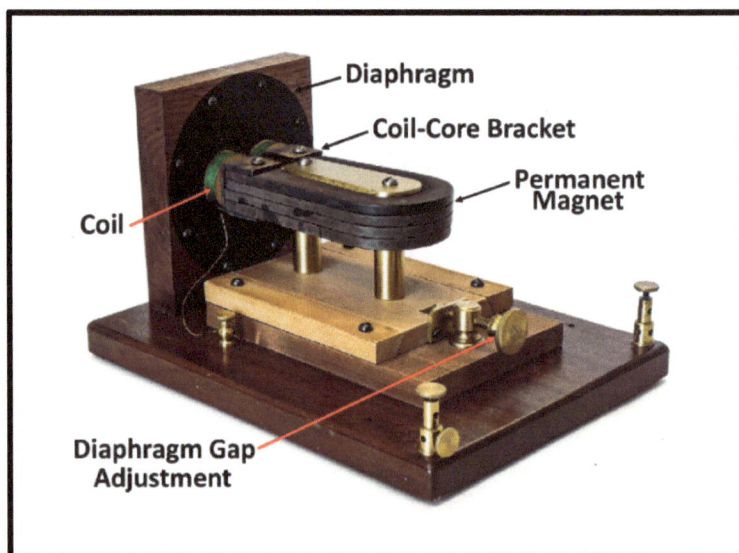

Fig. 12. Bell Labs replica of first commercial telephone.

By 1879, the young Bell System began introducing a less sophisticated, yet more powerful, transmitter that required batteries.

Bell's magneto telephone had a range of 100 miles, but this was not enough for long distance calls, so the magneto telephone was soon phased out as a transmitter. Although Bell's beautiful magneto transmitter is no longer found in common telephones, it is still used as a sound-powered transmitter on ships of the U.S. Navy, where it works even if the ship's power fails.

Receivers

The telephone in Fig. 12 also works in reverse, so to speak, as a receiver. In this case, an electric current such as just described flows through the coils and creates a time-varying magnetic field (Ampère's law). This magnetic field adds to and subtracts from the steady polarizing field from the permanent magnet, and the net result is a relatively strong magnetic field in the coil cores that increases and decreases a little with time. This varying field pulls on, and then relaxes, the iron diaphragm causing it to vibrate with the pitch and loudness of the sound at the transmitter.

The box telephone weighed a whopping 12 pounds, however, and a compact hand-held version of the magneto telephone was made that weighed less than a pound. Instead of using a horseshoe permanent magnet and two coils, the hand-held telephone used a permanent bar magnet and a single coil. A diagram of this telephone (Fig. 13) can be compared with Fig. 11 to see that the operating principle is the same.

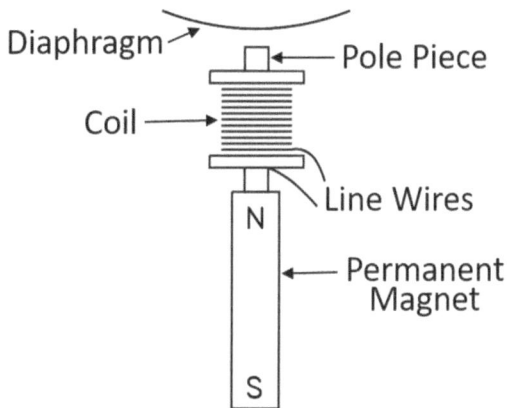

Fig. 13. Functional diagram of the hand-held magneto telephone.

The hand-held telephone could be used for talking or listening. It proved to be perfectly adequate as a receiver, but it was significantly less powerful than the box telephone as a transmitter. A Bell Labs replica of this hand telephone is shown in Fig. 14.

Fig. 14. Bell Labs replica of hand-held magneto telephone.

Of course you could use hand telephones and box telephones in any combination desired (Fig. 15). In practice, though, the Bell Company would install one box telephone and one hand telephone at each station. By mounting the box telephone on a wall as a transmitter and holding the small telephone to one's ear, conversations became more convenient. The hand-telephone design is still used as a receiver today, although even smaller components.

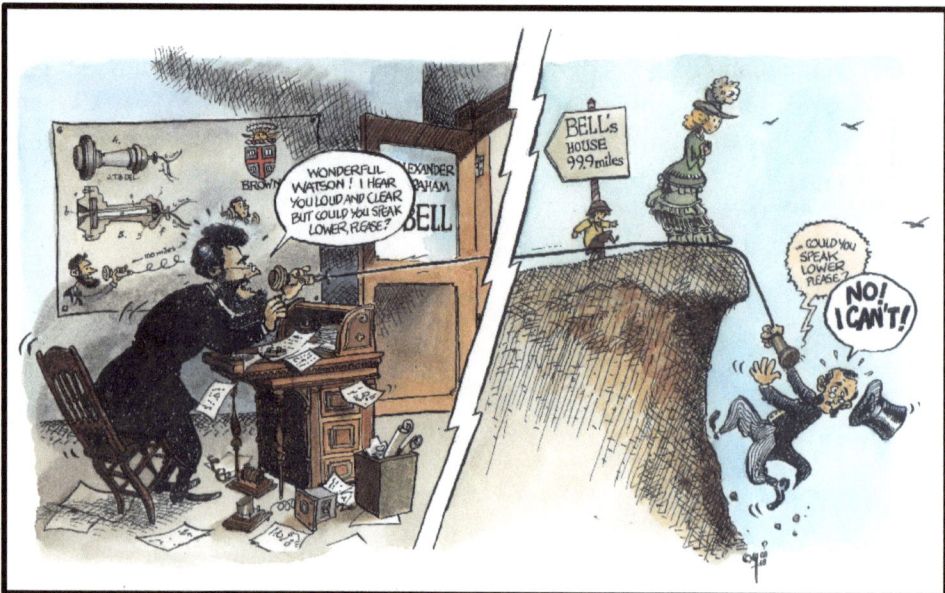

Fig. 15. Bell and Watson talking to each other on hand telephones.[3]

---

[3] Drawn by the French artist, Nicolas Waeckel, my friend and nuclear engineering colleague at Electricité de France.

Chapter 5. James Clerk Maxwell

James Clerk Maxwell, a Scotsman,[4] published a set of four seminal equations on electromagnetism in 1861 at the age of 30 – and the equations haven't been changed since.

Maxwell knew that a changing magnetic field would create an electric field, which in turn could push electrons around in a circuit (Chapter 3, Fig. 10). This was Faraday's law:

Rate of change of [Magnetic Field] → [Electric Field]

Maxwell also knew that current in an electric circuit would create a magnetic field near the circuit (Chapter 1, Fig. 2). This was Ampère's law:

[Electric Current] → [Magnetic Field]

There is a lot of symmetry in nature, and Maxwell thought there should be some here. Therefore, he postulated an additional term for Ampère's law that was analogous to Faraday's law. This term is known as Maxwell's addition:

Rate of change of [Electric Field] → [Magnetic Field]

Ampère's law with Maxwell's addition thus says that a magnetic field will be created by an electric current and also by an electric field that is changing with time.

Notice that a steady electric field in the circuit (e.g., from a battery) would cause a steady electric current. In that case, the rate of change of the electric field would be zero and Maxwell's additional term would be zero. Ampère's original law would then apply.

---

[4] The British are very good at electricity and magnetism, but they can't spell. His middle name was Clark.

However, as the rate of change of the electric field increases from zero, the rate-of-change term gets bigger. Thus high frequencies such as radio frequencies become very important.[5]

Maxwell had no experimental evidence, so his hypothesis was controversial. However, between 1886 and 1889, after more than 25 years of controversy, Heinrich Hertz, a German physicist, conducted experiments that proved Maxwell's addition was correct. By the way, most of these guys died young, and Maxwell did not live long enough to witness this.

Following is a rough representation of the two most important equations in Maxwell's set of equations. The other two equations are just details for our purposes.

[Electric Field] = d/dt [Magnetic Field]

[Magnetic Field] = [Electric Current] + d/dt [Electric Field]

Where d/dt indicates rate of change; i.e., the time derivative.

---

[5] Brit, my son (B.S. physics, M.S. electrical engineering), helped me substantially with clarity and accuracy.

Chapter 6. Radio

Wireless telegraphy and telephony are the holy grail of communication, and practical radio devices were developed by Marconi in the late 1890s. Imagine the importance of ship-to-shore communication during WWI (1914-1918), for example.

Radio Transmitters

In Chapter 1 (Fig. 2) we saw that a steady battery current (aka direct current) through a piece of wire in a circuit produces a steady magnetic field near that piece of wire. To be more precise, the battery produces a steady electric field in the wire, this field pushes electrons along steadily, and the moving electrical charges create the magnetic field. This was called Ampère's law. In fact, the steady magnetic field will exist everywhere in space – not just near the wire – but it doesn't appear instantly. It travels out from the wire at the speed of light, and the field strength falls off rapidly with distance from the wire. Figure 16 is a diagram of this situation, with a distant radio receiver on the right side.

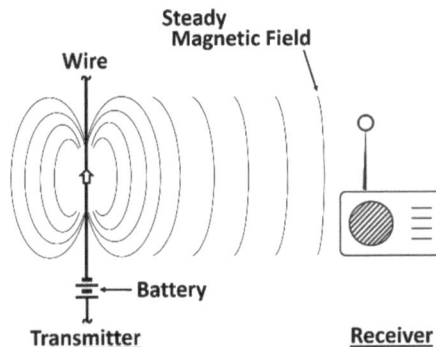

Fig. 16. Magnetic field lines from direct current.

In Chapter 3 (Fig. 9) we saw that a steady magnetic field does not produce current in a nearby wire. Therefore, when the magnetic field reaches the receiver (small delay), no current is produced in its antenna wire and the receiver doesn't detect a signal.

A radio transmitter's antenna is also a wire in a circuit, and a piece of this wire is depicted in Fig. 17. In a transmitter, a signal generator produces an electric field in the wire, and the strength of the field goes up and down in a cyclical fashion. This field, in turn, produces an alternating current in the antenna wire that increases and decreases with the same cyclical frequency.

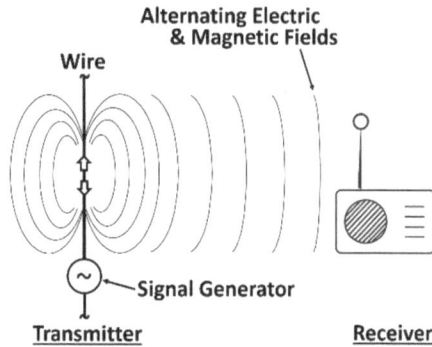

Fig. 17. Electric and magnetic field lines from alternating current.

Ampère's law with Maxwell's addition tells us that there are now two sources of a magnetic field that will exist outside the wire: one from the moving charges (the alternating current) and one from the changing electric field in the transmitter's antenna wire. In this case, the resulting magnetic field is changing with time, so it produces an electric field (Chapter 3, Fig. 10) that also exists outside the wire. The magnetic field and its companion electric field (i.e., the electromagnetic fields) travel out from the wire at the speed of light.[6] When they arrive (short delay), the electric field pushes electrons up and down in the receiver's antenna, and this is how the receiver detects a signal.

Utilization of radios followed the pattern set by telegraphs and telephones. At first, a telegraph key was put in the transmitter's circuit to turn the signal generator on and off in Morse-code fashion. In fact, this method is still used and is called continuous wave (CW) transmission or radio-telegraphy.

---

[6] Charles Evans, UNC professor of Physics, told me that the fields from the electric-current term decrease so rapidly with distance from the antenna that the stronger fields from Maxwell's addition make radio transmission practical.

Next came voice transmission. Sound pressure waves from speech go up and down in the range of 1 to 4 thousand times per second (1-4 kHz). Radios discussed here have signal generators that operate in the range of 540 to 1700 thousand times per second (540-1700 kHz or 0.54-1.70 MHz).[7] In the transmitter, sound pressure waves from speech (or music) activate a microphone, which is similar to a telephone transmitter, and the microphone produces an alternating current, let's say at 1 kHz (just below C_6 on a piano). With suitable amplification, this 1 kHz current is used to increase and decrease the strength of transmitter's signal generator, let's say at 1 MHz (middle of the radio dial). This is called amplitude modulation (AM) and it is literally multiplying two sinusoidal functions together, sin (1 kHz) times sin (1 MHz). Hence the transmitter's resulting 1 MHz electromagnetic fields are increasing and decreasing a thousand times per second (1 kHz). These modulated electromagnetic fields are called radio waves. Audio-frequency modulation enables audible sounds to be detected by a radio receiver. More said below.

Radio Receivers

During WWI (1914-1918), amateur radio enthusiasts were forbidden to transmit or listen to radio communications. Frank Conrad, a Westinghouse engineer, had a special license to test the performance of radio equipment after work between his garage and the Westinghouse laboratory 5 miles away in Pittsburgh. Conrad had a habit of putting his microphone in front of a Victrola record player to get uninterrupted sound for his tests. As soon as the ban was lifted, word spread about Conrad's "evening concerts," which amateurs apparently had been listening to all along.

Within a year, Westinghouse realized that radios could be used for entertainment as well as communication, and then there would be a big market for selling radio receivers. This concept of broadcasting entertainment programs was new and was first put into operation on Westinghouse's radio station KDKA in Pittsburgh.

The first radio receiver that was produced for this entertainment market was the Westinghouse *Aeriola Jr.* crystal radio that required

---

[7] This is the AM band on a regular radio. The FM band is at 88-108 MHz, and there are many other special bands above and below these frequencies.

no external power. It simply used the power in its antenna current to make headphones (just sensitive telephone receivers) work. *Aeriola Jr.* went on sale in July 1921 and I bought one (Fig. 18) on eBay in March 2022.

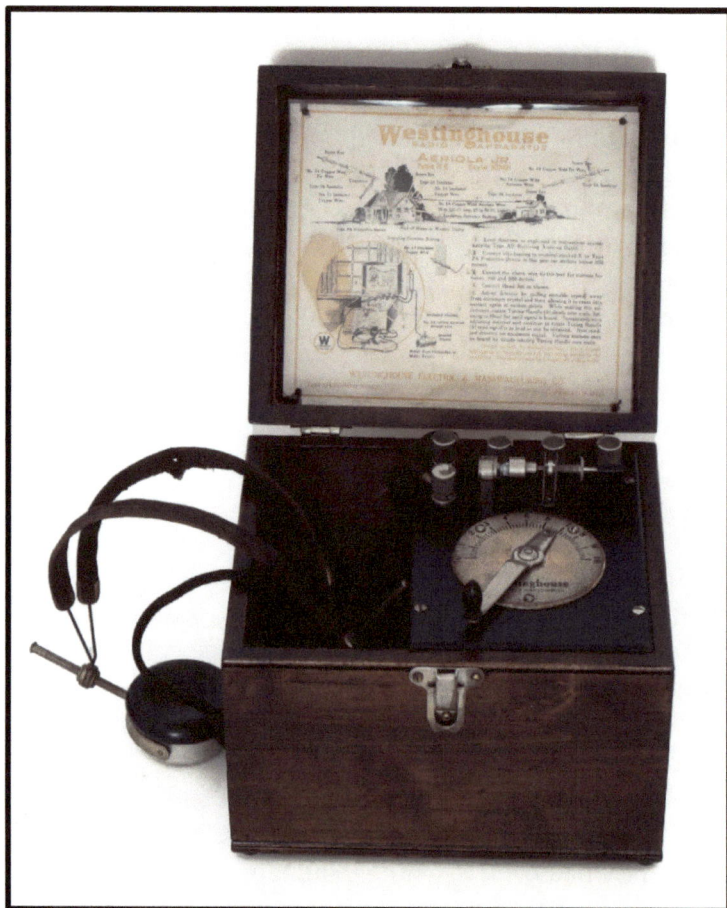

Fig. 18. Westinghouse *Aeriola Jr.* crystal radio receiver of 1921.

From the discussion above, you can guess that the antenna for *Aeriola Jr.* is going to be very important. The antenna is just a wire, so it's necessary to dig a little deeper into the electrical properties of a wire. It is also necessary to recognize that the ground (i.e., the earth) affects the antenna wire. The ground is a good electrical conductor, so imagine the earth as a big metal ball. Radios always use the earth to complete their circuits (charges cannot be created or destroyed; they are just pushed around in closed circuits).

Three different properties of a wire can impede the flow of charges through wire in a radio circuit: resistance (R), inductance (L), and capacitance (C). Separate components are also made with specific values of R, L, and C for use in circuit design, and they are called resistors, inductors, and capacitors, respectively.

We've already mentioned that a wire contains electrons that can be pushed around easily by an electric field – easily, but not freely. A wire will have some impedance to current flow that is related to its atomic structure, and this property is called resistance. It is the same for direct current and alternating current, and the symbol for a resistor (-W-) will be used below to represent a wire's resistance.

Above, in the discussion around Fig. 17, we saw that an electric field is created outside a wire that is conducting alternating current. Well, that field is also felt inside the wire and it pushes back on the flowing electrons. In other words, the electric field that was created by an alternating current impedes the current that created it. A wire's length and shape determine its inductance, and the symbol for an inductor (~) will be used below to represent a wire's inductive impedance. Inductive impedance increases with the frequency of alternating current and is zero for steady current.

A capacitor is usually constructed with two conducting surfaces (such as tin foil) that are separated by an insulating material (such as paper). Thus there is a gap between the conductors, and electrons can't go across the gap. The symbol for a capacitor (-||-) will be used below to represent a wire's capacitive impedance, which decreases with frequency and is infinite for steady current.

If a capacitor is inserted in a circuit with a battery, a steady current cannot be established. Some charges will be pushed up to the gap, where they accumulate, but the charges can't get across the gap. This is just like an open circuit.

For a circuit with alternating current, however, the charges that accumulate on one plate soon turn around and go back to the other plate. That way, if the signal generator is alternating, there is a lot of back-and-forth charge motion. This motion is in fact an alternating current, although no charges ever get across the gap.

We will now look in detail at the *Aeriola Jr.* because its circuit is the basis for radio circuits that followed.

Radio-Frequency Loop

The instructions on the label in the *Aeriola Jr.* radio call for an antenna to be constructed from a horizontal wire that's 75 to 150 feet long and 25 to 50 feet high.[8] The electrical properties of a long antenna wire can now be discussed. An antenna is often represented by an inverted triangle in circuit diagrams, so that's the meaning of the faint triangle in Fig. 19.

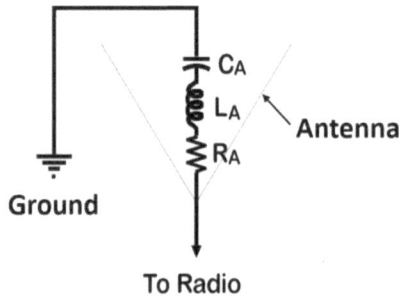

Fig. 19. Equivalent circuit of antenna wire.

All pieces of wire will have some inherent resistance; thus a representation of the antenna will include a resistance, R_A. The subscript, A, will be used to indicate that these properties belong to the antenna.

Signal generators in radio transmitters operate at very high frequencies, so a long straight wire will have an inductive impedance that is significant. Thus the antenna representation will include an inductance, L_A.

Here comes the spooky part. The antenna wire is long, and although it is 25-50 feet above the ground, the wire acts as one plate of a

---

[8] Because the transmitter's antenna is a long vertical wire (tall tower), the electric field that arrives at the receiver is aligned in the vertical direction. Hence the 25 to 50-foot vertical length of the receiver's antenna wire is where the electric field pushes charges up and down. The remaining horizontal length has to do with wave reflections and capacitance (details beyond the scope of *Naked Physics*).

capacitor while the earth's surface acts as the other. Yes, the capacitance is very small, but the frequency is very high, so the impedance is not infinite. Therefore, the antenna representation in Fig. 19 also includes a capacitor, C_A, and this capacitor connects the tip end of the antenna wire to ground!

Radio Frequency Loop

The basic circuit for the *Aeriola Jr.* receiver is shown Fig. 20.[9] On the left side, the antenna's equivalent circuit is connected to ground just as in Fig. 19. Below the antenna, the signal generator symbol now represents the electric field that has traveled from the transmitter and is pushing electrons up and down the receiver's antenna wire. This is the driver, so to speak, for all the current in the receiver. Everything below the signal generator symbol in Fig. 20 is inside the *Aeriola Jr.* set.

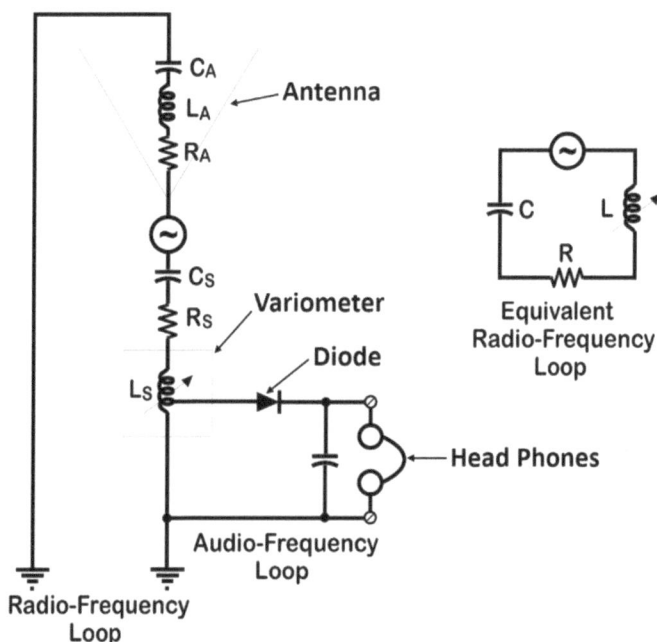

Fig. 20. Circuit diagrams for *Aeriola Jr.* with its antenna.

***

[9] I had help understanding this from Eric Wenaas (Ph.D.), author of *Radiola: The Golden Age of RCA.*

An extra capacitor C_S (S for set) is located inside the *Aeriola Jr.*'s set, and this extra capacitance is needed for tuning (see below). A variable inductor, L_S, called a Variometer is also inside the set for tuning. The wires inside the set also have some resistance, R_S. The wires also have some inductance and capacitance, but we will just consider those to be included in L_S and C_S. Finally, the *Aeriola Jr.* set is connected to ground.

Thus the main part of the diagram is a radio-frequency loop, and it is a completely closed circuit! Who would have thought? For 70 years I could not understand how a radio worked without having a complete circuit. That loose end of the antenna wire is actually connected to ground by its capacitance – and hence it's also connected to the receiver via the ground. By the way, the same is true of the transmitter's antenna, and its capacitance completes the circuit referred to in Fig. 17.

The separate resistances, inductances and capacitances in the diagram of the radio-frequency loop can now be added together and represented by R, L, and C, respectively. These additions produce the simplified equivalent diagram on the right in Fig. 20. This simplified diagram is a classic "tank circuit" that has interesting properties and is used for tuning.

As the incoming electric field (represented by the signal generator symbol) tries to push electrons around the tank circuit, the current is impeded by a resistance, an inductance, and a capacitance. Keep in mind that the inductive impedance increases as frequency increases and that the capacitive impedance decreases as frequency increases. There is, therefore, a happy medium – a frequency at which the total impedance around the radio frequency loop is a minimum such that the current is maximized. This occurs at a special frequency that depends on the values of L and C (details).

This circuit provides selectivity such that the radio receiver responds well at the special frequency and poorly at other frequencies. Selecting this special frequency is called tuning. In *Aeriola Jr.*, the tuning frequency is altered by changing the inductance of the variable inductor with a dial selector.

Audio Frequency Loop

Now consider the audio-frequency loop in Fig. 20. Headphones cannot respond at radio frequencies because their diaphragms are too massive to vibrate that fast. You couldn't hear those frequencies anyway. Thus it's necessary to obtain the envelope of the current oscillations as the envelope goes up and down at audio frequencies. This process is depicted in Fig. 21.

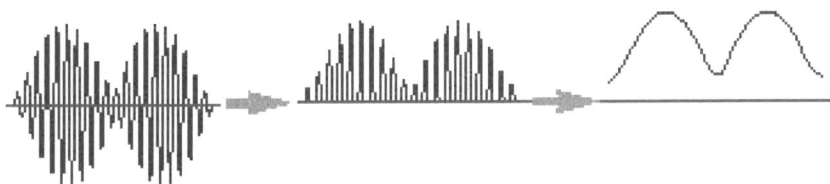

Fig. 21. Current in various parts of the audio-frequency loop.

The graph on the left represents current that is going up and down at a radio frequency in the shared portion of the audio-frequency and radio-frequency loops. The magnitude of this current is responding to changes in electric field strength, which is being modulated by the transmitter at the audio frequency and amplitude of the sound input. To get the audio envelope of this radio-frequency current, the negative portions of the current are first prevented from entering the audio-frequency loop, and this requires a "detector."

The *Aeriola Jr.* has a Perikon crystal detector (Fig. 22) that uses two crystals to form a primitive semiconductor diode. One crystal is bornite and the other is zincite. A diode has a low resistance in one direction that lets current pass, but it has a high resistance in the other direction that prevents current from passing. Thus the graph in the center in Fig. 21 shows that only positive current gets through the crystal diode; no negative current passes. This current is still changing at a radio frequency, though, and can't be heard in the headphones.

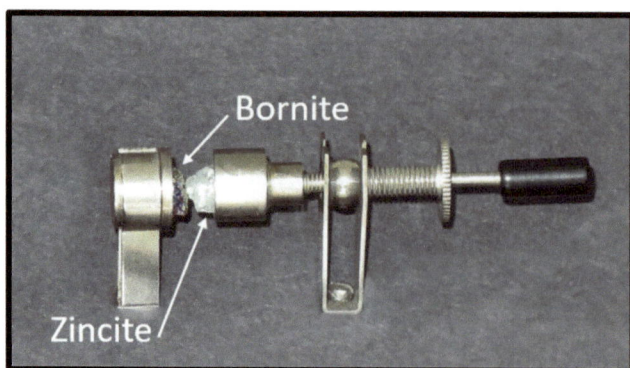

Fig. 22. Perikon crystal detector in *Aeriola Jr.*

The capacitor in the audio-frequency loop finishes the job. As the positive current emerges from the diode in one of the radio-frequency oscillations, some of electric charges accumulate in the capacitor. After the current from the diode has peaked and is decreasing to stay at zero for half a cycle, charges begins to flow back out of the capacitor, which acts literally as a reservoir for electric charges. Thus current continues to flow through the headphones during that period between two radio-frequency peaks. This detail is not shown in the graphs in Fig. 21, but if it had been shown it would appear as a little saw-tooth ripple in the audio-frequency result on the right-hand side of Fig. 21. The massive diaphragms in the headphones are insensitive to this radio-frequency ripple.

Finally, there is the mater of attaching the audio-frequency loop to the radio-frequency loop. One could attach the audio-frequency loop at the top of the inductor, but the goal is to deliver as much power to the headphones as possible. This happens when impedances are matched, not necessarily at the top of the inductor, and this engineering detail is beyond the scope of *Naked Physics*.

---------- *The End* ----------

About the back cover:

The big transmitter tower for the WSM radio station was erected in 1932 and is still being used today. It was first of the Blaw-Knox Co. diamond-shaped antennas, and the metal mast structure itself functions as the antenna – in effect a long vertical wire. The antenna is 808-feet tall (246 meters), which is effectively half the wavelength of WSM's 650 kHz broadcast frequency. The tower is stabilized by guy wires, and the point at the bottom rests on a large porcelain insulator. WSM is the home of The Grand Ole Opry in Nashville. Two years after the WSM tower was built, Blaw-Knox Co. built a similar tower for the WLW radio station in Cincinnati (one of the author's childhood local stations). The WLW tower is slightly smaller, corresponding to the somewhat higher broadcast frequency of 700 kHz, and it also remains in operation today.

Porcelain Insulator

Porcelain insulator on bottom of WSM radio tower.

About the author:

Ralph O. Meyer is a Phi Beta Kappa graduate of the University of Kentucky and earned a Ph.D. in physics from the University of North Carolina. His primary technical work was with the U.S. Nuclear Regulatory Commission. However, for more than 30 years, he has studied and written about the technical history and development of the telephone. For the past few years he has been working with Associate Professor Russell A. Flinchum at North Carolina State University on telephone-related design topics.

www.ingramcontent.com/pod-product-compliance
Lightning Source LLC
Chambersburg PA
CBHW041720200326

41520CB00005B/223